LABORATORY NOTEBOOK

P. A. Kramer

© 2015 P. A. Kramer

ISBN 978-1-312-90198-8

THE
LABORATORY NOTEBOOK
OF

_____

NOTEBOOK

\# _____

DATE

_____ / _____ / _____ — _____ / _____ / _____

PROPERTY OF THE LABORATORY OF

_____

AT

_____

TABLE OF CONTENTS

#	EXPERIMENT	DATE	PAGE #
-	Instructions for use	MM / DD / YY	ix
		/ /	
		/ /	
		/ /	
		/ /	
		/ /	
		/ /	
		/ /	
		/ /	
		/ /	
		/ /	
		/ /	
		/ /	
		/ /	
		/ /	
		/ /	
		/ /	
		/ /	
		/ DD / YY	
		/ /	
		/ /	
		/ /	
		/ /	
		/ /	
		/ /	
		/ /	
		/ /	
		/ /	
		/ /	
		/ /	

#	EXPERIMENT	DATE	PAGE #
		/ /	
		/ /	
		/ /	
		/ /	
		/ /	
		/ /	
		/ /	
		/ /	
		/ /	
		/ /	
		/ /	
		/ /	
		/ /	
		/ /	
		/ /	
		/ /	
		/ /	
		/ /	
		/ /	
		/ /	
		/ /	
		/ /	
		/ /	
		/ /	
		/ /	
		/ /	
		/ /	
		/ /	
		/ /	
#	EXPERIMENT	DATE /	PAGE #

#	EXPERIMENT	DATE	PAGE #
		/ /	
		/ /	
		/ /	
		/ /	
		/ /	
		/ /	
		/ /	
		/ /	
		/ /	
		/ /	
		/ /	
		/ /	
		/ /	
		/ /	
		/ /	
		/ /	
		/ /	
		/ /	
		/ /	
		/ /	
		/ /	
		/ /	
		/ /	
		/ /	
		/ /	
		/ /	
		/ /	
		/ /	
		/ /	
		/ /	
		/ /	

#	EXPERIMENT	DATE	PAGE #
		/ /	
		/ /	
		/ /	
		/ /	
		/ /	
		/ /	
		/ /	
		/ /	
		/ /	
		/ /	
		/ /	
		/ /	
		/ /	
		/ /	
		/ /	
		/ /	
		/ /	
		/ /	
		/ /	
		/ /	
		/ /	
		/ /	
		/ /	
		/ /	
		/ /	
		/ /	
		/ /	
		/ /	
		/ /	

Instructions for use:

1. Write legibly using a ball-point pen that will not smear.

2. Enter Experiment Title, Number, and Date (MM/DD/YY), and indicate if it is a continuation of a previous experiment.

3. In the provided box, draw an experimental flowchart or scheme for easy reference.

4. Record the experimental objective, observations, and summary.

5. Maintain careful record of experimental protocols, techniques, instrumentation, and reagents associated with each experiment.

6. Show calculations and record steps systematically.

7. Obtain signature from supervisor or primary investigator if necessary.

8. Do not remove pages.

9. Do not alter the contents of an experiment after its completion without careful notation (initials and date).

10. If electronic content exists for an experiment, note the file name and location.

11. Populate the TABLE OF CONTENTS during or after completion of the LABORATORY NOTEBOOK.

x

EXPERIMENT TITLE _____ .

EXPERIMENT # _____ ☐ CONTINUED   DATE _____ / ____ / ____

EXPERIMENT TITLE _____

EXPERIMENT # _____ ☐ CONTINUED   DATE ____ / ____ / ____

4

EXPERIMENT TITLE _____ .

EXPERIMENT # _____ ☐ CONTINUED   DATE _____ / _____ / _____

EXPERIMENT TITLE _____ :

EXPERIMENT # _____ □ CONTINUED   DATE ____ / ____ / ____

8

EXPERIMENT TITLE _____.

EXPERIMENT # _____ □ CONTINUED   DATE _____ / ____ / ____

EXPERIMENT TITLE _____

EXPERIMENT # _____ ☐ CONTINUED   DATE ____ / ____ / ____

EXPERIMENT TITLE _____ .

EXPERIMENT # _____ ☐ CONTINUED   DATE _____ / _____ / _____

13

Performed by _____   Reviewed by _____

EXPERIMENT TITLE _____ .

EXPERIMENT # _____ □ CONTINUED   DATE _____ / ____ / _____

17

19

23

EXPERIMENT TITLE _____ .

EXPERIMENT # _____ ☐ CONTINUED   DATE ____ / ____ / ____

25

29

EXPERIMENT TITLE _____ .

EXPERIMENT # _____ □ CONTINUED   DATE ____ / ____ / ____

EXPERIMENT TITLE _____

EXPERIMENT # _____ □ CONTINUED   DATE _____ / ___ / ___

35

EXPERIMENT TITLE _____ .

EXPERIMENT # _____ □ CONTINUED   DATE ____ / ____ / _____

41

43

45

47

EXPERIMENT TITLE _____ .

EXPERIMENT # _____ ☐ CONTINUED   DATE _____ / _____ / _____

EXPERIMENT TITLE _____

EXPERIMENT # _____ ☐ CONTINUED   DATE _____ / _____ / _____

51

EXPERIMENT TITLE _____

EXPERIMENT # _____ ☐ CONTINUED   DATE ____ / ____ / ____

EXPERIMENT TITLE _____ .

EXPERIMENT # _____ ☐ CONTINUED DATE ____ / ____ / ____

55

Performed by _____   Reviewed by _____

EXPERIMENT TITLE _____

EXPERIMENT # _____ ☐ CONTINUED   DATE ____ / ____ / ____

EXPERIMENT TITLE

EXPERIMENT # _____ ☐ CONTINUED   DATE ____ / ____ / ____

63

63

65

67

EXPERIMENT TITLE _____ .

EXPERIMENT # _____ ☐ CONTINUED   DATE ____ / ____ / ____

73

74

75

EXPERIMENT TITLE _____ .

EXPERIMENT # _____ ☐ CONTINUED   DATE _____ / _____ / _____

81

83

85

86

Performed by _____ Reviewed by _____

Performed by _____ Reviewed by _____

89

EXPERIMENT TITLE _____ .

EXPERIMENT # _____ ☐ CONTINUED   DATE _____ / _____ / _____

EXPERIMENT TITLE _____

EXPERIMENT # _____ ☐ CONTINUED   DATE _____ / _____ / _____

95

EXPERIMENT TITLE _____ .

EXPERIMENT # _____ ☐ CONTINUED   DATE _____ / _____ / _____

99

EXPERIMENT TITLE _____ .

EXPERIMENT # _____ ☐ CONTINUED   DATE ____ / ____ / ____

EXPERIMENT TITLE _____ .

EXPERIMENT # _____ ☐ CONTINUED   DATE _____ / ____ / ____

Performed by _____ Reviewed by _____

118

EXPERIMENT TITLE _____ .

EXPERIMENT # _____ □ CONTINUED   DATE ____ / ____ / ____

Performed by _____ Reviewed by _____

122

EXPERIMENT TITLE _____

EXPERIMENT # _____ ☐ CONTINUED   DATE _____ / _____ / _____

EXPERIMENT TITLE _____

EXPERIMENT # _____ ☐ CONTINUED   DATE ____ / ____ / ____

EXPERIMENT TITLE _____ .

EXPERIMENT # _____ ☐ CONTINUED   DATE ____ / ____ / ____

EXPERIMENT TITLE _____

EXPERIMENT # _____ ☐ CONTINUED   DATE ____ / ____ / ____

135

EXPERIMENT TITLE _____.

EXPERIMENT # _____ ☐ CONTINUED   DATE _____ / ____ / _____

EXPERIMENT TITLE _____ .

EXPERIMENT # _____ ☐ CONTINUED   DATE ____ / ___ / ____

EXPERIMENT # _____ ☐ CONTINUED   DATE ____ / ___ / ____

139

EXPERIMENT TITLE _____

EXPERIMENT # _____ ☐ CONTINUED   DATE ____ / ____ / ____

EXPERIMENT TITLE _____

EXPERIMENT # _____ ☐ CONTINUED   DATE _____ / _____ / _____

146

EXPERIMENT TITLE _____ .

EXPERIMENT # _____ □ CONTINUED   DATE ____ / ____ / _____

EXPERIMENT TITLE _____ .

EXPERIMENT # _____ ☐ CONTINUED   DATE ____ / ____ / ____

153

155

EXPERIMENT TITLE _____

EXPERIMENT # _____ ☐ CONTINUED   DATE _____ / ___ / _____

EXPERIMENT TITLE _____

EXPERIMENT # _____ □ CONTINUED   DATE ____ / ____ / ____

Performed by _____ Reviewed by _____

162

EXPERIMENT TITLE _____

EXPERIMENT # _____ ☐ CONTINUED   DATE _____ / _____ / _____

EXPERIMENT TITLE _____

EXPERIMENT # _____ ☐ CONTINUED   DATE ___ / ___ / ___

EXPERIMENT TITLE _____ .

EXPERIMENT # _____ ☐ CONTINUED   DATE _____ / _____ / _____

EXPERIMENT TITLE _____

EXPERIMENT # _____ ☐ CONTINUED   DATE ____ / ____ / ____

EXPERIMENT TITLE

EXPERIMENT # _____ ☐ CONTINUED   DATE ____ / ____ / ____

www.ingramcontent.com/pod-product-compliance
Lightning Source LLC
Chambersburg PA
CBHW081046170526
45158CB00006B/1868